Pre-Algebra

Test Booklet

1-888-854-MATH (6284)
www.MathUSee.com

1-888-854-MATH (6284)
www.MathUSee.com

In other words, thou shalt not steal.

Printed in the United States of America

TEST

1

Fill in the blank with "have" or "owe."

1. + 19 = ______19

2. -35 = ______35

3. -58 = ______ 58

Add.

4. (+8) + (-7) =

5. (-10) + (-2) =

6. (-7) + (-15) =

7. (+9) + (-11) =

8. (+32) + (+96) =

9. (+4) + (-13) =

10. (-5) + (-18) =

11. (-436) + (-251) =

12. (-511) + (+709) =

Find the fraction of the number.

13. $\frac{1}{5}$ of 10 =

14. $\frac{2}{3}$ of 9 =

15. $\frac{7}{8}$ of 32 =

16. $\frac{1}{2}$ of 30 =

17. Austin took 25 paces forward. Then he took 17 paces backward. How many paces from his starting point was he?

18. Greg owes $5 to one friend and $6 to another. Express his debt as a negative number.

19. Four-fifths of the people at the party won prizes. If there were 20 people at the party, how many won prizes?

20. The book Devan is reading has 254 pages. If Devan is one-half of the way through the book, how many pages has he read?

Change to an addition problem, then solve.

1. (+52) – (–23) =

2. (–35) – (–16) =

3. (+54) – (+15) =

4. (–7) – (+ 24) =

5. (–36) – (+ 49) =

6. (+22) – (–30) =

Change the signs as needed and solve.

7. (+30) + (–24) =

8. (–53) – (+10) =

9. (+13) + (–2) =

10. (–33) – (+2) =

11. (–7) + (+1) =

12. (–4) – (+18) =

Add or subtract.

13. $\frac{1}{5} + \frac{3}{5}$

14. $\frac{2}{8} - \frac{1}{8} =$

15. $\frac{5}{7} - \frac{2}{7} =$

16. $\frac{1}{3} + \frac{1}{3} =$

17. Two-fourths of the cake was eaten on Monday and one-fourth was eaten on Tuesday. What part of the cake has been eaten?

18. Nick has 15 dimes. If he gives Sean one-third of the dimes, how many will Nick have left?

Write your answers as positive or negative numbers.

19. Buddy received $21 and spent $25. What was the total effect on his budget?

20. Jake found 46 coins on the beach with his metal detector but lost 15 out of his pocket later that day. How many coins did Jake have left?

TEST

3

Multiply.

1. $(-20) \times (-4) =$
2. $(+19) \times (-3) =$
3. $(-30) \times (-17) =$
4. $(-27) \times (+8) =$
5. $(-9) \times (+2) =$
6. $(-7) \times (-29) =$

Change the signs as needed and solve.

7. $(+33) - (-46) =$
8. $(-27) + (-10) =$
9. $(-41) - (-20) =$

Find the fraction of the number.

10. $\frac{1}{3}$ of 24 =
11. $\frac{2}{5}$ of 15 =
12. $\frac{3}{7}$ of 28 =

Add or subtract. Leave answers in the form in which they occur.

13. $\frac{5}{8} - \frac{3}{8} =$
14. $\frac{7}{10} - \frac{1}{10} =$
15. $\frac{1}{4} + \frac{1}{4} =$

Fill in the missing numbers to make equivalent fractions.

16. $\frac{1}{5} = \frac{\quad}{\quad} = \frac{\quad}{15} = \frac{4}{\quad}$

17. $\frac{2}{3} = \frac{\quad}{\quad} = \frac{\quad}{\quad} = \frac{8}{12}$

18. Emily did one-fifth of the chores and Madison did three-fifths of them. What part of the chores has been done?

Write your answers as positive or negative numbers.

19. Elizabeth spent $4 a day on lunch for five days. Write the daily cost of the lunch as a negative number and multiply to find the total change in the amount of money she has.

20. During the drought, the water level in the lake fell two feet (–2) every week. What was the effect on the water level in six weeks?

TEST 4

Solve each problem as indicated.

1. $\frac{-6}{-2} =$

2. $35 \div (-7) =$

3. $-6 \overline{)\ -48}$

4. $\frac{-28}{-4} =$

5. $26 \div (-2) =$

6. $7 \overline{)\ -56}$

7. $(-3) \times (+6) =$

8. $(+4) \times (-2) =$

9. $(-5) \times (-6) =$

10. $(-38) + (+12) =$

11. $(+47) + (-39) =$

12. $(-21) - (-45) =$

Add or subtract. Reduce answers to lowest terms.

13. $\frac{1}{8} + \frac{3}{8} =$

14. $\frac{4}{10} + \frac{4}{10} =$

15. $\frac{7}{8} - \frac{1}{8} =$

16. Graph (-3) and (+3), then find the distance between them by counting the spaces.

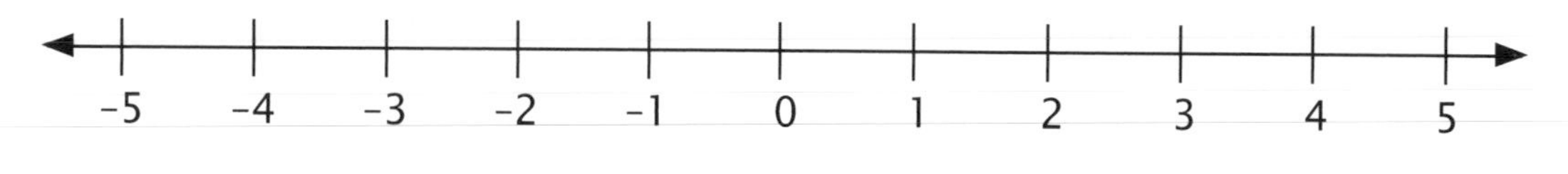

D = ____

17. Is the number (-7) a whole number? Is it an integer?

18. Dustin got four-fifths of the answers correct on his math test. If there were 20 questions, how many did he get right?

19. One-third of the birds in my yard were woodpeckers and one-third were chickadees. What part of the birds was either woodpeckers or chickadees?

20. Allen has a debt of $500. He owes equal amounts to 10 different people. Express his debt to one person as a negative number.

TEST

5

Rewrite each number without an exponent.

1. $5^2 =$

2. $1^6 =$

3. $3^3 =$

4. $\left(\frac{1}{2}\right)^4 =$

Write the missing exponent.

5. $15 \times 15 \times 15 \times 15 = 15^{\_\_}$

6. $10^{\_\_} = 1{,}000$

7. $32 = 2^{\_\_}$

8. $4 \times 4 \times 4 \times 4 \times 4 = 4^{\_\_}$

9. $5^{\_\_} = 125$

10. $49 = 7^{\_\_}$

Write the missing number in the numerator to make each pair of fractions equivalent.

11. $\frac{1}{3} = \frac{\_\_}{21}$

12. $\frac{3}{4} = \frac{\_\_}{16}$

13. $\frac{7}{10} = \frac{\_\_}{20}$

Use the rule of four to compare the fractions, then tell which is the largest.

14. $\frac{1}{3} \bigcirc \frac{3}{8}$

15. $\frac{2}{5} \bigcirc \frac{4}{7}$

16. $\frac{3}{4} \bigcirc \frac{5}{9}$

17. Write the following three different ways: three used as a factor four times.

18. Ariana had 12 new pencils. If she gave 3/4 of them away, how many did she give away?

    How many pencils did Ariana have left?

19. Jane has a debt of $63. She owes equal amounts to seven different people. Express her debt to one person as a negative number.

20. Nina spent $12 on a shirt and $7 on a scarf. What was the total effect on her budget?

Write in exponential notation.

1. 2.247 =

2. 156.3 =

3. 3,910 =

4. 78.09 =

Write in standard notation.

5. $5 \times 10^3 + 4 \times 10^2 + 1 \times 10^1 + 7 \times 10^0 + 4 \times \frac{1}{10^1} + 1 \times \frac{1}{10^2} + 2 \times \frac{1}{10^3} =$

6. $4 \times 10^2 + 1 \times \frac{1}{10^1} + 3 \times \frac{1}{10^2} =$

Fill in the blanks.

7. $4.02 = 4 ______ and 0 ______ and 2 ______ =

______ + ______ + ______ = ______

8. $8.15 = ______ dollars and ______ dimes and ______ pennies =

______ + ______ + ______ = ______

Write the missing exponent.

9. $2^{\_\_} = 8$

10. $10^{\_\_} = 100$

Solve each problem as indicated.

11. (-33) + (-22) =

12. (-26)(10) =

13. (91) ÷ (-13) =

14. (19) - (-18) =

Add or subtract. Write your answer in simplest form.

15. $\frac{2}{3}+\frac{1}{6}=$

16. $\frac{5}{8}-\frac{1}{4}=$

17. $\frac{1}{7}+\frac{5}{6}=$

18. Two-thirds of the people in the group came in blue cars, and one-eighth of them came in red cars. If there were 48 people in the group, how many came in red or blue cars?

19. Abigail bought 3/4 of a pound of hard candy. If she has 3/8 of a pound of candy left, how much has she eaten?

20. Brent has six dimes, five pennies, and eight dollars. Express his total amount in decimal form.

Simplify.

1. $-(5)^2 =$

2. $-7^2 =$

3. $(-10)^3 =$

4. $\left(-\frac{4}{9}\right)^2 =$

5. $-2^3 =$

6. $(-3)^2 =$

7. $-(6)^2 =$

8. $-\left(\frac{1}{8}\right)^2 =$

Write in exponential notation.

9. 6.19 =

Write in standard notation.

10. $4 \times 10^2 + 9 \times 10^1 + 8 \times \frac{1}{10^1} + 2 \times \frac{1}{10^2} + 3 \times \frac{1}{10^3} =$

Solve each problem as indicated.

11. $(-23) + (36) =$

12. $(-8)(-28) =$

13. $(-22) \div (-2) =$

Add or subtract. Write your answers in lowest terms.

14. $\frac{1}{3} + \frac{2}{5} =$

15. $\frac{5}{9} + \frac{1}{4} =$

16. $\frac{3}{4} - \frac{2}{3} =$

Multiply.

17. $\frac{3}{8} \times \frac{1}{6} =$

18. $\frac{1}{2} \times \frac{5}{8} =$

19. $\frac{11}{12} \times \frac{3}{7} =$

20. Kara promised to do 5/9 of the job. So far she has done only 3/5 of what she promised to do. What part of the whole job has Kara finished?

UNIT TEST **Lessons 1-7**

Add, subtract, multiply, or divide.

1. $(-8) + (-17) =$

2. $(-9) \times (35) =$

3. $(13) - (-21) =$

4. $(-36) \div (-6) =$

5. $(-10) \times (-10) =$

6. $(5) + (-7) =$

7. $(-72) \div (9) =$

8. $(-2) - (14) =$

9. $(50) \div (-5) =$

10. $(-38) - (-12) =$

11. $(-46) + (61) =$

12. $(7) \times (-91) =$

Simplify.

13. $-2^3 =$

14. $-(4)^2 =$

15. $(-10)^2 =$

16. $\left(-\frac{5}{8}\right)^2 =$

Write in exponential notation.

17. 165.9 =

18. 4.038 =

Write in standard notation.

19. $7 \times 10^3 + 3 \times 10^2 + 9 \times \frac{1}{10^1} + 1 \times \frac{1}{10^2} + 4 \times \frac{1}{10^3} =$

Add or subtract. Write your answers in lowest terms.

20. $\frac{1}{7} + \frac{2}{5} =$

21. $\frac{3}{10} + \frac{1}{5} =$

22. $\frac{7}{8} - \frac{3}{4} =$

Multiply and write your answers in lowest terms.

23. $\frac{7}{8} \times \frac{1}{4} =$

24. $\frac{1}{2} \times \frac{5}{9} =$

25. $\frac{2}{3} \times \frac{11}{12} =$

Find the fraction of the number.

26. $\frac{1}{3}$ of 36 =

27. $\frac{2}{5}$ of 40 =

28. $\frac{3}{7}$ of 21 =

TEST

8

Solve.

1. $\sqrt{16} =$
2. $\sqrt{9} =$
3. $\sqrt{7^2} =$
4. $\sqrt{A^2} =$
5. $\sqrt{100} =$
6. $\sqrt{81} =$
7. $\sqrt{11^2} =$
8. $\sqrt{X^2} =$

Simplify.

9. $-3^3 =$
10. $(-2)^2 =$
11. $\left(\frac{1}{100}\right)^2 =$
12. $-\left(\frac{4}{5}\right)^2 =$

Solve each problem as indicated.

13. $(5) + (-12) =$
14. $(-21) \times (-15) =$
15. $(36) \div (-3) =$

Divide. Use the rule of four to make the denominators the same. Reduce only if the result is a whole number.

16. $\frac{1}{2} \div \frac{1}{4} =$

17. $\frac{7}{8} \div \frac{4}{5} =$

18. $\frac{5}{8} \div \frac{1}{10} =$

19. Six-eighths of the pie is left in the pan. Peggy wants to divide the leftover pie among her guests so that each can take home one-fourth of a pie. How many people can have one-fourth of a pie?

20. If a square has an area of 49 square feet, what is the length of each side of the square?

TEST

9

Solve for the unknown by using the additive inverse. Check your work by substituting the answer in the original equation.

1. $X - 16 = 42$
2. Check

3. $-X + 4 = -2X - 6$
4. Check

5. $4R - 4 = 3R + 10$
6. Check

7. $2Y - 3 = Y - 4$
8. Check

Simplify each expression.

9. $\sqrt{25^2} =$
10. $\sqrt{64} =$

11. $-(3)^2 =$
12. $\left(\frac{1}{2}\right)^3 =$

Divide using the rule of four. Reduce fractions, but do not change to mixed numbers. Simplify fractions when the result is a whole number.

13. $\frac{1}{4} \div \frac{1}{8} =$

14. $\frac{2}{3} \div \frac{1}{2} =$

15. $\frac{5}{12} \div \frac{3}{4} =$

Write the reciprocal of each number, then multiply to check.

16. $\frac{5}{6}$

17. 32

18. $\frac{1}{9}$

19. Ryan is X years old. Two times his age plus fifteen equals thirty-seven minus two. Write an equation showing how old Ryan is. Solve it if you can.

20. Andy is two-fifths of Ruth's age. If Ruth is ten, how old is Andy?

TEST

10

Use the Pythagorean theorem to answer the questions.

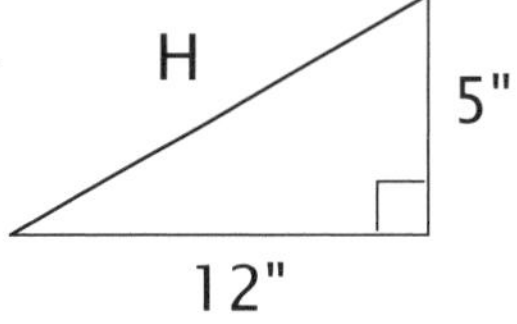

1. What is the length of the hypotenuse?

2. Can a right triangle have sides of 12, 16, and 20?

3. What is the length of the unknown side?

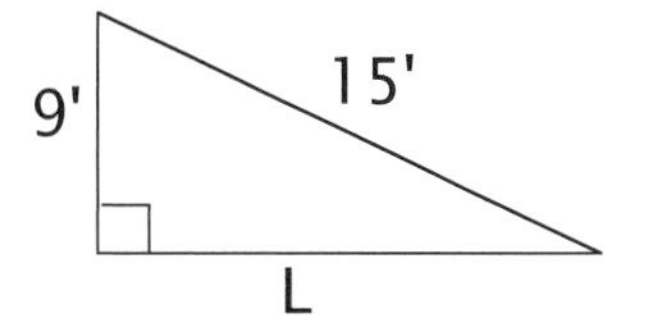

4. Can a right triangle have sides of 5, 10, and 15?

Solve for the unknown using the additive inverse.

5. $A - 7 = -13$

6. Check

7. $10X - 8 = 9X + 8$

8. Check

Simplify each expression.

9. $\sqrt{X^2} =$

10. $(-8)^2 =$

11. $\left(\frac{1}{2}\right)^3 =$

12. $\sqrt{\frac{9}{16}} =$

Divide using the short cut. Reduce fractions, but do not change to mixed numbers. Simplify fractions when the result is a whole number.

13. $\frac{3}{8} \div \frac{1}{6} =$

14. $\frac{5}{9} \div \frac{1}{4} =$

15. $\frac{4}{5} \div \frac{3}{5} =$

16. Solve: (4) – (–10) =

17. Write in standard notation:

$$3\text{x}10^2 + 5\text{x}10^1 + 8\text{x}10^0 + 1\text{x}\frac{1}{10^2}$$

18. Nimrod wanted to build a tower three miles high. How far would it be from the top of his tower to a point four miles away from the base of the tower?

19. During the drought, the water level in the lake fell two feet (–2) each week. What was the effect on the water level in six weeks? Write your answer as a positive or negative number.

20. Two times a number, plus five, equals the number, minus five. Write an equation and find the number.

TEST

11

Fill in the blanks.

1. The associative property is true for ______________ and ______________.

2. The associative property is false for ______________ and ______________.

3. The commutative property is true for ______________ and ______________.

4. The commutative property is false for ______________ and ______________ .

Use the associative and commutative properties to simplify by combining like terms.

5. $-5X + 8X - 4$

6. $-5Y + 3 - 6Y + 2Y + 4$

7. $6 + X - 5 + 2X + 8$

8. $3B - B + 7 + 4B$

Simplify by combining terms, then solve for the unknown and check.

9. $4A - 1 = 3A + 8$

10. Check

11. $1 + F + 3 + F = F + 5$

12. Check

Multiply or divide. Change any improper fractions to mixed numbers and reduce answers as possible.

13. $\frac{1}{4} \times \frac{4}{5} =$

14. $\frac{7}{10} \div \frac{1}{2} =$

15. $\frac{3}{4} \div \frac{9}{10} =$

16. The hypotenuse of a right triangle is 15 feet, and one leg is 9 feet. What is the length of the other leg?

17. What is the square root of 81?

18. Naomi had B bushels of grain. After Ruth brought her 5 more bushels of grain, she had 11 bushels in all. How much grain did Naomi have to start with? Write an equation and solve.

19. Two-thirds of Sandi's rose bushes bloomed this summer. One-half of the ones that bloomed were pink. What part of Sandi's total rose bushes had pink blooms this summer?

    If Sandi had 12 rose bushes, how many bore pink blooms?

20. Mom has three-quarters of a pound of chocolates left. If she divides the chocolates into one-eighth-pound portions and eats one portion a day, how many days will the chocolates last her?

Rewrite each expression using the distributive property.

1. $5(3X + 5Y) =$

2. $B(7 + 8) =$

3. $X(4C + 3D) =$

4. $9(Q + W) =$

5. $XY(2 + 2) =$

6. $2Y(5A + 6B) =$

Find the common factor and rewrite.

7. $10X + 10Y =$

8. $8A + 11A =$

9. $4XY + 8XZ =$

10. $4Q + 4R =$

11. $5A + 25B =$

12. $AD + CD =$

Solve for the unknown. Check your answer by substituting it in the original equation.

13. $-A - 22 = -2A - 30$

14. Check

15. $Y - 12 = 2Y - 4$

16. Check

Give the length of each line as sixteenths of an inch. Reduce if possible.

17. 0" 1"

18. 0" 1"

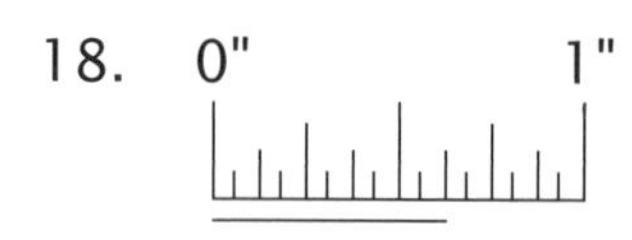

19. 0" 1"

20. Derrick discovered that seven times X plus five times X was the same as eleven times X, minus fifteen. Write an equation and solve to find the value of X.

TEST 13

Simplify and solve for the unknown. Use the additive and multiplicative inverses.

1. $4X = -124$
2. Check

3. $-7Q + 6 + 5Q = 15 - 7$
4. Check

5. $3(P + 5) + P = 3(2 + P)$
6. Check

7. $2(A + 4) + 6A = 2(2 + 3A)$
8. Check

Give the length of each line as sixteenths of an inch. Reduce if possible.

9. 0" 1"

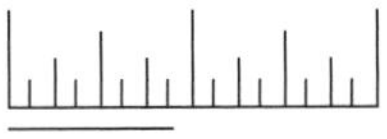

10. 0" 1"

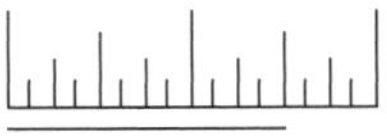

11. 0" 1"

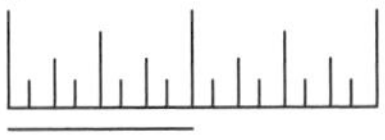

Multiply or divide.

12. $\frac{7}{8} \times \frac{2}{7} =$

13. $\frac{4}{5} \div \frac{1}{9} =$

14. $\frac{5}{6} \div \frac{4}{12} =$

Add.

15. $1\frac{3}{5} + 2\frac{3}{4}$

16. $3\frac{1}{6} + 5\frac{2}{7}$

17. $9\frac{1}{2} + 4\frac{1}{2}$

18. Two times a number, plus eight, equals sixteen. Write an equation and solve to find the number.

19. A farmer bought a field in the shape of a right triangle. The hypotenuse was five miles long and one leg was four miles long. How long was the road running along the third side?

20. When you multiply by the multiplicative inverse, your answer should always be what number?

TEST

14

Simplify these expressions using PARAchute EXpert My Dear Aunt Sally.

1. $7^2 + 2^2 - 5 - 4 + 3X$

2. $X + 32 \div 4 - 2^2$

3. $-Y - 5 + Y + 2(2Y - Y) - 3$

4. $5X - 3 - X - 3(X + 1^2)$

Simplify and solve for the unknown. Use order of operations as needed. Check your work.

5. $5(B + 3) = 4(B - 7) + 2B$

6. Check

7. $5^3 - 10^2 = X(8 - 2) + 2X - 3X$

8. Check

9. $(-3)^2 + (8 + 3^2) = 2A$

10. Check

Simplify.

11. $-(8)^2 =$

12. $2^3 =$

13. $-4^2 =$

14. $1^5 =$

Subtract. Be sure your answer is reduced.

15. $7\frac{1}{4} - 1\frac{3}{8}$

16. $9 - 2\frac{2}{3}$

17. $10\frac{1}{3} - 6\frac{5}{9}$

18. We got 3 1/2 inches of snow last week and 5 3/4 inches of snow today. How much more snow fell today than last week?

19. Three times a number, minus nine, equals six times six, divided by four. Write an equation and find the number.

20. Seventy-two people bought ice cream at Tom's store. Three-fourths of them bought chocolate ice cream. How many did **not** buy chocolate ice cream?

UNIT TEST **Lessons 8-14**

Simplify each expression.

1. $\sqrt{36} =$

2. $\sqrt{R^2} =$

3. $\sqrt{64} =$

4. $\frac{3}{\sqrt{25}} =$

Simplify these expressions using PARAchute EXpert My Dear Aunt Sally.

5. $4^2 - 3(5 - 2) - 25 + 6$

6. $13 + 49 \div 7 - 2^2$

7. $(3 \times 6^2 - 1) + 11$

8. $3(20 - 4^2) + 2 \times 3$

Simplify and solve for the unknown. Use order of operations as needed. Check your work.

9. $3 + 10 - R + 6R = -3 + 9R + 5 - 5$

10. Check

11. $(-3)^2 + (F + 3^2) = 2 \times 4 + 6$

12. Check

13. $-3X + 4X = 2 \cdot 4 - X$

14. Check

Give the length of each line as sixteenths of an inch. Reduce if possible.

15. 0" 1"

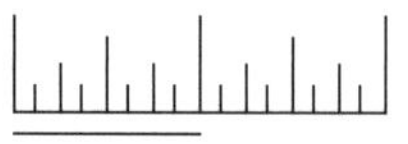

16. 0" 1"

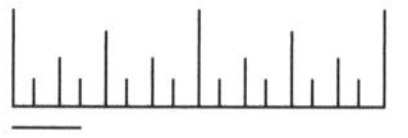

17. 0" 1"

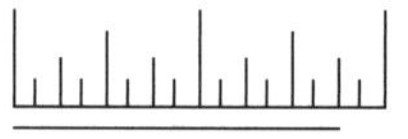

Use the Pythagorean theorem to solve for the unknown side.

18.

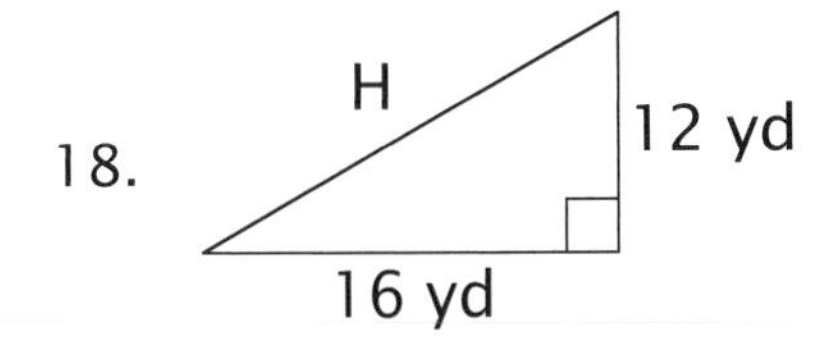

19.

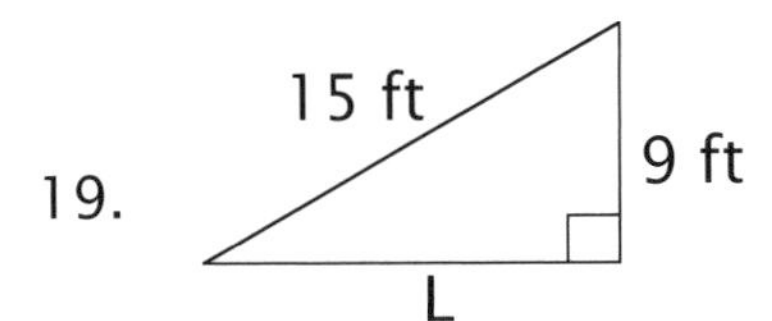

Divide. Change any improper fractions to mixed numbers.

20. $\frac{1}{2} \div \frac{1}{6} =$

21. $\frac{7}{8} \div \frac{3}{4} =$

22. $\frac{5}{7} \div \frac{5}{9} =$

Add or subtract. Be sure your answer is reduced.

23. $3\frac{5}{8} + 2\frac{5}{6}$

24. $8 - 1\frac{1}{3}$

25. $$\begin{array}{r} 9\frac{1}{8} \\ -\ 5\frac{4}{5} \\ \hline \end{array}$$

26. A triangle has sides of five feet, six feet, and eight feet. Is it a right triangle?

27. Eight times a number, minus five, equals seven times the number, plus five. Write an equation and find the number.

28. Is division commutative?

TEST 15

Find the surface area.

1. 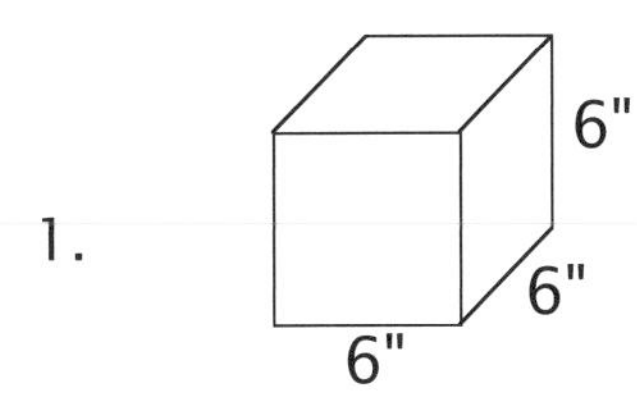

2. 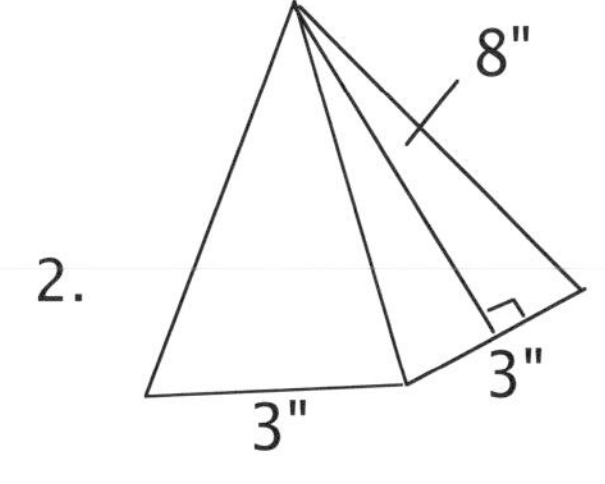

3. 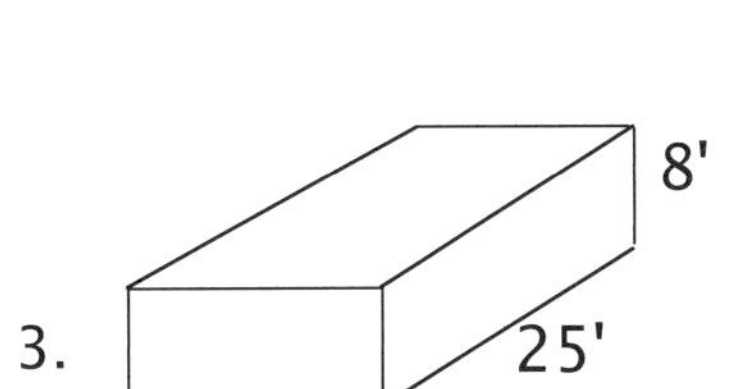

4. 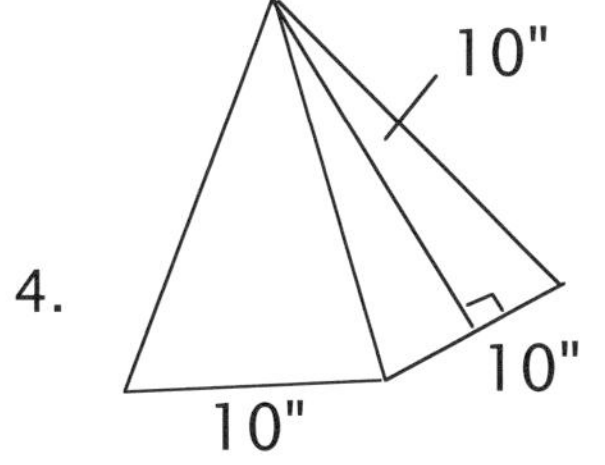

Simplify and solve for the unknown. Check your work.

5. $5B + 10 = -10$

6. Check

7. $A - 9 = -14$

8. Check

9. $4(J + 4) + 3J = 3(28 + J)$

10. Check

11. $(-5)^2 + (9 + 4^2) = 5B$

12. Check

Multiply.

13. $2\frac{7}{8} \times 2\frac{6}{9} \times 1\frac{1}{2} =$

14. $3\frac{1}{4} \times 6\frac{2}{3} \times 5\frac{1}{10} =$

15. $\frac{1}{7} \times 1\frac{2}{5} \times 4\frac{1}{5} =$

16. What is the area of the roof of the house? (Remember, there are two slanted sides to the roof.)

17. Since a square of shingles is 100 square feet, how many squares are needed to cover the roof (see #16)? You cannot buy a fraction of a square.

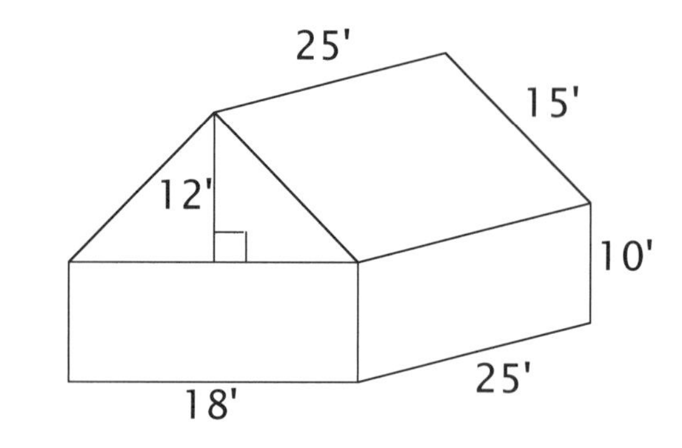

18. Find the surface area of the sides of the house. Include the two triangular sections.

19. A square of vinyl siding is also 100 square feet. Using your answer for #18, tell how many squares of vinyl siding you would need to buy to cover the sides of the house. You cannot buy a fraction of a square.

20. Bryan had 24 coins in his pocket. He lost one-third of his coins. One-half of the lost coins were pennies. How many pennies did he lose?

TEST

16

Use the formula to change Celsius to Fahrenheit.

1. $60^\circ C =$ _____ F
2. $110^\circ C =$ _____ F
3. $13^\circ C =$ _____ F
4. $42^\circ C =$ _____ F

Simplify and solve for the unknown. Check your answers.

5. $Q + 4 = 3Q - 6$
6. Check
7. $7^2 - X - 1 = 5X$
8. Check

Add or subtract.

9. $\begin{array}{r} 7.00 \\ +\ .45 \\ \hline \end{array}$
10. $\begin{array}{r} 11.2 \\ -\ 3 \\ \hline \end{array}$
11. $\begin{array}{r} 6.15 \\ +\ 2.2 \\ \hline \end{array}$
12. $\begin{array}{r} 58.9 \\ -\ 7.1 \\ \hline \end{array}$

13. Write the following in exponential notation: 61.32

14. Give the freezing point of water in Celsius and in Fahrenheit.

15. Give the boiling point of water in Celsius and in Fahrenheit.

16. Give normal body temperature in Celsius and in Fahrenheit.

17. A rectangular solid has a base that measures 12 feet by 16 feet, and a height of two feet. What is its surface area?

18. A pyramid with a square base measures one yard on a side. The height of each face is two yards. What is the surface area of the pyramid?

19. Hailey walked 2.32 miles in the morning and 1.5 miles in the afternoon. How many miles did she walk in all?

20. Hannah walked 1 3/4 miles in the morning and 3 2/3 miles in the afternoon. How many miles did Hannah walk in all?

TEST

17

Use the formula to change Fahrenheit to Celsius or Celsius to Fahrenheit as indicated.

1. $41^\circ F =$ _____ C

2. $140^\circ F =$ _____ C

3. $91^\circ F =$ _____ C

4. $200^\circ C =$ _____ F

5. $35^\circ C =$ _____ F

6. $56^\circ C =$ _____ F

Simplify and solve for the unknown. Check your work.

7. $3(5) + 1 = -5X + X$

8. Check

9. $D(2 - 5) - 8 = -3D - 2D + 6$

10. Check

Multiply.

11. $\begin{array}{r} 2.8 \\ \times\ .31 \\ \hline \end{array}$

12. $\begin{array}{r} .456 \\ \times\ \ .2 \\ \hline \end{array}$

13. $\begin{array}{r} .78 \\ \times\ .59 \\ \hline \end{array}$

14. $1\frac{1}{6} \times 5\frac{3}{8} =$

15. $2\frac{4}{5} \times 2\frac{1}{2} =$

16. $3\frac{1}{10} \times 3\frac{1}{3} =$

17. Andrew bought a sandwich for $4.37. How much change should he get from a $10 bill?

18. A book cost $8.98. How much money does Kaitlyn need to buy three books?

19. The weather report said the temperature was 31 °C. Is it above or below freezing?

20. Ashley's rectangular bedroom measures 12 feet by 11 feet. The ceiling is 8 feet high. Ashley wants to put wallpaper on the walls. What is the surface area of the four walls of her room?

TEST

18

Simplify each expression.

1. $(8 - 4)^2 \times |4 - 8| =$

2. $11 + |3 + 2^2| =$

3. $|1 - 3^2| + |-5| =$

Simplify and solve for the unknown.

4. $2Y - 2 = 3Y - |6 - 12|$

5. Check

6. $-2X + |6 + 1| + 3X - 4 = 10 - 1$

7. Check

8. $4A + 5 - 3 = 2A + (2)^2(2)$

9. Check

10. $-2B + 3 + 5B + 1^2 = 2(3 + 2) + 3^2$

11. Check

Use the formula to change Fahrenheit to Celsius or Celsius to Fahrenheit as indicated.

12. $35^\circ C =$ _____ F

13. $72^\circ F =$ _____ C

Divide. Add zeros as necessary to divide without remainder.

14. $5\overline{)4.35}$

15. $8\overline{)23.04}$

16. $4\overline{)18}$

17. $3\overline{)2.16}$

18. What is the boiling point of water in Celsius?

19. What is the surface area of a cube that measures 5.2 inches on each side?

20. A cook decided that 3.6 pounds of meat was enough to buy. If he needs to serve nine people, how many pounds of meat per person is the cook planning on?

Solve for the unknown using whichever method is most convenient.

1. $\frac{7}{12} = \frac{A}{60}$

2. $\frac{9}{10} = \frac{81}{T}$

3. $\frac{6}{G} = \frac{8}{16}$

4. $\frac{3}{5} = \frac{Q}{25}$

5. $\frac{6}{11} = \frac{36}{X}$

6. $\frac{2}{D} = \frac{3}{9}$

Simplify and solve for the unknown.

7. $2(2X + 1) = 2(X + 4)$

8. Check

9. $11 + Q = |(-3^2)| + 4^2 - 5$

10. Check

Add, subtract, or multiply as indicated.

11. $1.8 \times .23 =$ 12. $.169 + 1.5 =$ 13. $17.2 - 3.4 =$

Divide.

14. $.5\overline{)1.5}$ 15. $.08\overline{)96}$ 16. $.06\overline{).12}$

17. Last month we had only five sunny days. If there were 30 days last month, what was the ratio of sunny days to the total number of days?

    What was the ratio of sunny days to days that were not sunny?

18. Three-fifths of the days last year were sunny. If there were 365 days in the year, how many were sunny?

19. A rope 75 meters long was cut into pieces with a length of 1.5 meters each. How many pieces resulted?

20. Thirty-two people entered the room. One-half of them decided to stay and listen to the speaker. Of those who stayed to listen, three-quarters were bored. How many of those who stayed were not bored?

TEST

20

Write a proportion for each set of similar polygons and solve for the unknown side.

1.

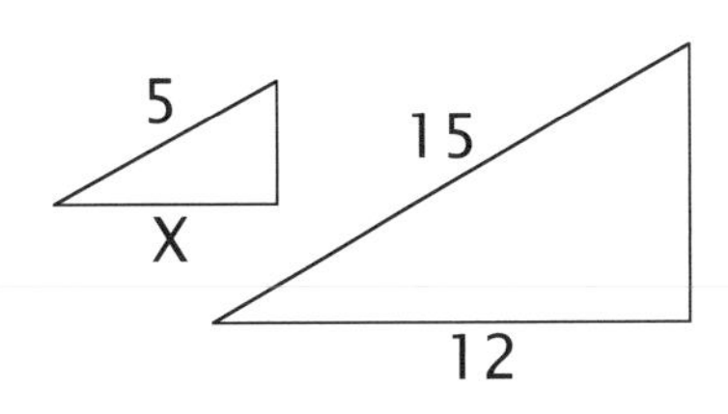

2.

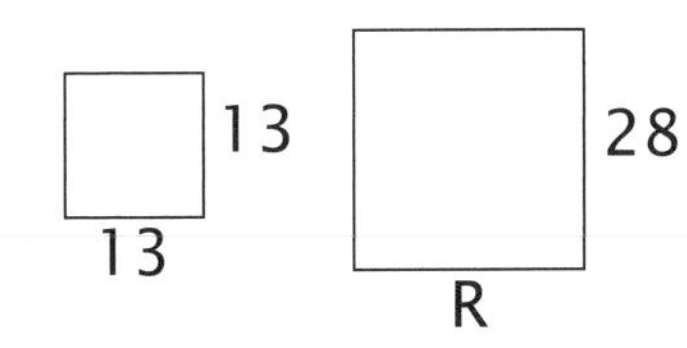

3.

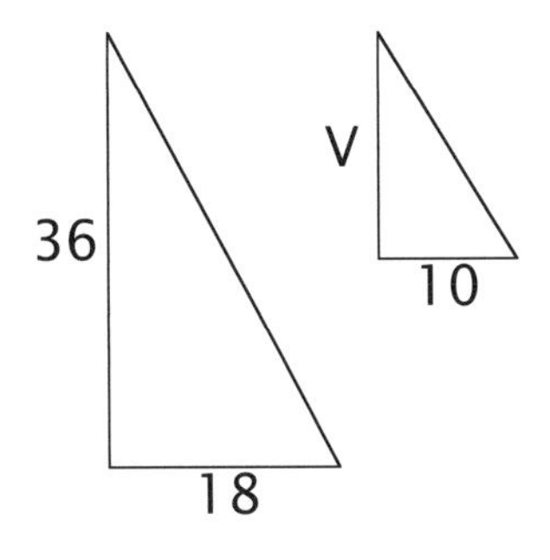

4.

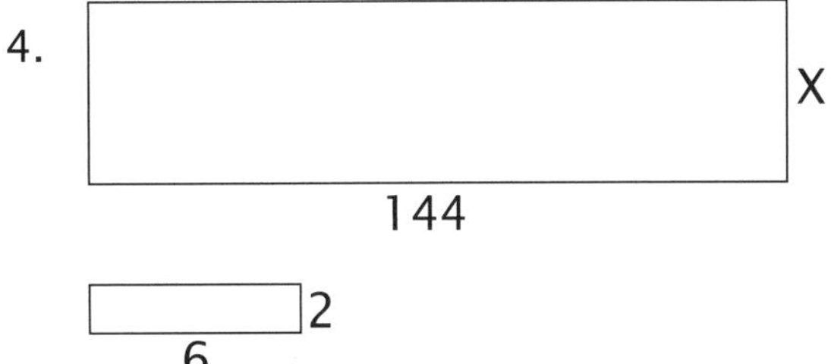

Solve for the unknown.

5. $\frac{5}{6} = \frac{45}{Y}$

6. $\frac{9}{G} = \frac{63}{70}$

Change the fractions to decimals.

7. $\frac{3}{8}$

8. $\frac{10}{100}$

9. $\frac{5}{10}$

10. $\frac{1}{4}$

Change the decimals to reduced fractions.

11. .45

12. .8

13. .66

14. .4

Simplify and solve for the unknown.

15. $3X - 9 + 7X - 10 = 9X - 5X + 5$

16. Check

17. $5^2 \div 5 + 3(X + 7) = 2X + 27$

18. Check

19. Brian ate .75 of a pie. Tell what part of the pie was left over with a decimal, and with a fraction.

20. What is the difference in height between a tree that is 40.5 feet tall and one that is 36.2 feet tall?

TEST

21

Find the prime factors for each number.

1. 18
2. 25
3. 7

Find the LCM for each pair of numbers. Use whichever method you prefer.

4. 32 and 64
5. 5 and 7
6. 25 and 10
7. 11 and 12
8. 8 and 12
9. 4 and 6

Write a proportion for each set of similar polygons and solve for the unknown side.

10. 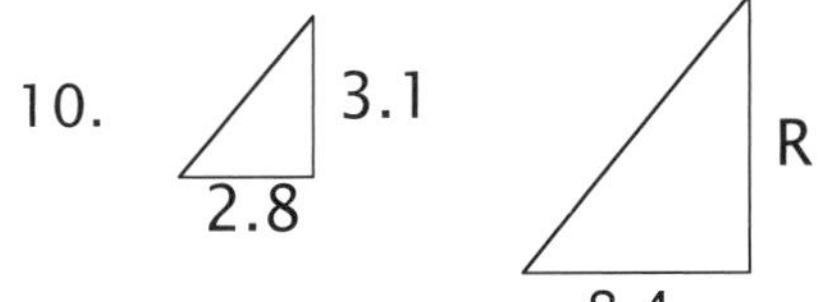

11. 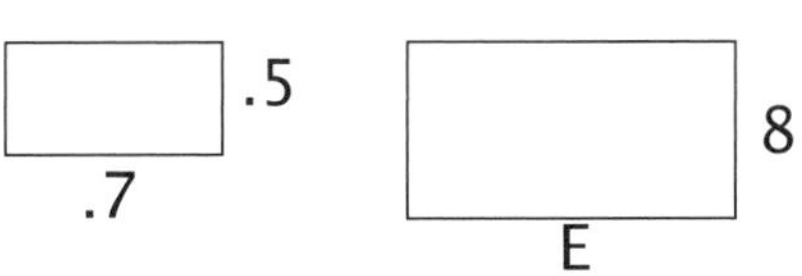

Write each fraction as a decimal and as a percent. Divide decimals to the hundredths place and write any remainders as fractions.

12. $\frac{1}{4}$ = =

13. $\frac{2}{3}$ = =

14. $\frac{3}{100} = \qquad =$

15. $\frac{7}{8} = \qquad =$

16. $\frac{1}{2} = \qquad =$

17. $\frac{40}{100} = \qquad =$

18. A triangle has sides of three, five, and seven. Is it a right triangle?

19. Tony picked 5 1/2 bushels of tomatoes. He gave 1 3/8 bushels to his grandmother and sold the rest. How many bushels did he sell?

20. Ian's mom sang him S songs, then his sister sang him two times as many as his mom had. If Ian listened to nine songs in all, how many did his mom sing? Write an equation and solve.

TEST

22

Find the GCF for each pair of numbers.

1. 24 and 42
2. 21 and 14
3. 44 and 36
4. 12 and 48
5. 3 and 9
6. 10 and 60

Find the LCM for each pair of numbers.

7. 3 and 6
8. 8 and 12
9. 5 and 15

Follow the signs.

10. (-13) + (-45) =
11. (-32) x 19 =
12. -1 - 2 =

Find the percent of each number.

13. 35% of 100 =
14. 2% of 4.6 =
15. 10% of .95 =

Solve for the unknown. Check your answers.

16. $25 \div 5 + 3(X + 7) = 2X + 3^3$

17. $4X + 3^2 - 9 + 17 = 27 - X$

18. Greg had $50 in his wallet. If he spent $12.45 in the grocery store and $35.95 in the bookstore, how much money did he have left?

19. Fifty items were lost. Ashley found 20% of them on Monday and 10% of them on Tuesday. How many items are still to be found?

20. Mom said that Bret could have one-fourth of the pie. Express that amount as a decimal and as a percent.

UNIT TEST Lessons 15-22

Solve for the unknown.

1. $\frac{1}{3} = \frac{25}{Y}$

2. $\frac{14}{32} = \frac{A}{16}$

3. $\frac{7}{G} = \frac{7}{8}$

Write a proportion for each set of similar polygons and solve for the unknown side.

4.

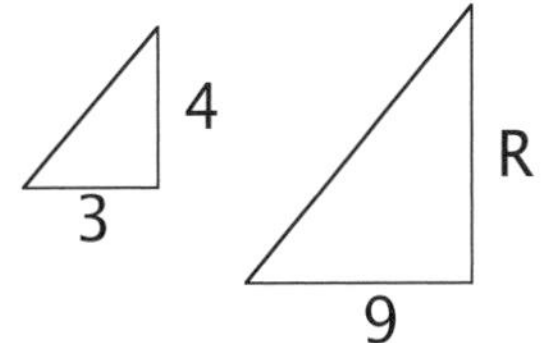

5.

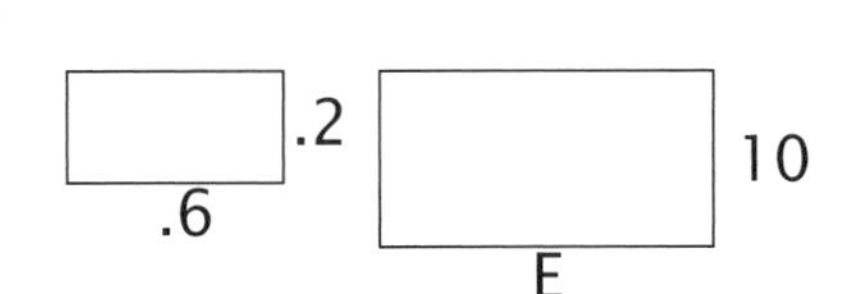

Find the LCM for each pair of numbers.

6. 4 and 5

7. 15 and 20

8. 12 and 32

Find the GCF for each pair of numbers.

9. 11 and 22

10. 16 and 18

11. 10 and 25

12. A room is 15 feet long, 13 feet wide, and 10 feet high. How many square feet must be covered if the walls, ceiling, and floor are all to be painted the same color?

13. What is the surface area of a pyramid with a square base measuring 10 feet on a side, if the height of each face is five feet?

14. The temperature in Nigel's back yard is 30 ° Celsius. Give the temperature in Fahrenheit and tell if it is more likely to be winter or summer.

15. The weather report gave the temperature as 41 ° Fahrenheit. What is the Celsius temperature?

16. Simplify and write without the absolute value signs:

$|2 - 4^2| + |5| =$

Add or subtract.

17. $\begin{array}{r} 8.5 \\ +\ 2.0 \\ \hline \end{array}$

18. $\begin{array}{r} 5.28 \\ -\ 3.04 \\ \hline \end{array}$

19. $\begin{array}{r} 4. \\ +\ 2.69 \\ \hline \end{array}$

20. $\begin{array}{r} 6.0 \\ -\ 1.7 \\ \hline \end{array}$

Multiply.

21. $\begin{array}{r} 5.9 \\ \times\ .4 \\ \hline \end{array}$

22. $\begin{array}{r} .006 \\ \times\ .36 \\ \hline \end{array}$

23. $\begin{array}{r} 7.8 \\ \times\ 3.1 \\ \hline \end{array}$

Divide.

24. $7\overline{)2.8}$

25. $.05\overline{)75}$

26. $.03\overline{).06}$

Write each fraction as a decimal and as a percent. Divide decimals to the hundredths place and write any remainders as fractions.

27. $\frac{4}{5} =$ $=$

28. $\frac{1}{3} =$ $=$

29. $\frac{16}{100} =$ $=$

30. Kate plans to save 5% of her income. If she just earned $80, how much should she save?

Add the polynomials.

1. $\begin{array}{r} 7X^2 - 4X - 1 \\ -\ 5X^2 + 2X + 4 \\ \hline \end{array}$

2. $\begin{array}{r} 3X^2 + 3X + 6 \\ +1X^2 + 6X - 1 \\ \hline \end{array}$

3. $\begin{array}{r} -X^2 - 2X - 2 \\ -\ 4X^2 + X + 9 \\ \hline \end{array}$

4. $\begin{array}{r} 9X^2 - 6X - 3 \\ -6X^2 + 3X + 8 \\ \hline \end{array}$

5. $\begin{array}{r} 5X^2 + 4X + 7 \\ +3X^2 + 8X - 2 \\ \hline \end{array}$

6. $\begin{array}{r} -7X^2 - X - 5 \\ -\ X^2 +\ 5X + 1 \\ \hline \end{array}$

Solve for the unknown.

7. $\frac{2}{5} = \frac{R}{100}$

8. $\frac{46}{X} = \frac{2}{7}$

9. $\frac{11}{50} = \frac{22}{Q}$

Find the percent of each number.

10. 500% of 3.4 =

11. 250% of 78.2 =

12. 120% of 40 =

13. The LCM of 6 and 8 = ____ .

14. The GCF of 6 and 21 = ____.

Find the surface area.

15.

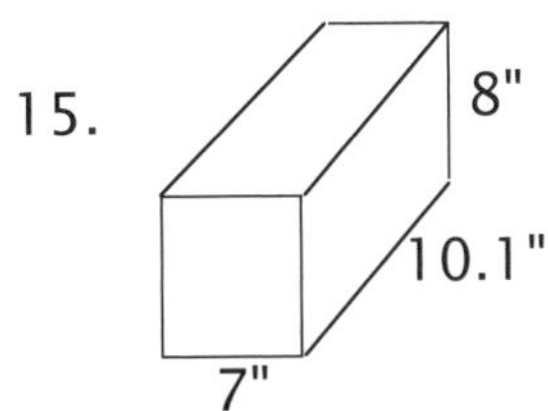

16.

.12'
.5'
.5'

17. What is the boiling point of water in Fahrenheit?

18. Janet bought 3/4 of a pound of apples, 1 1/2 pounds of bananas, and 2 1/4 pounds of grapes. How many pounds of fruit did she buy?

19. The walk-a-thon raised 170% of the expected amount. If the goal was $1,500.00, how much was raised?

20. Three times a number, plus nine, equals negative two times the same number, plus forty-four. What is the number?

TEST

24

Find the volume.

1. 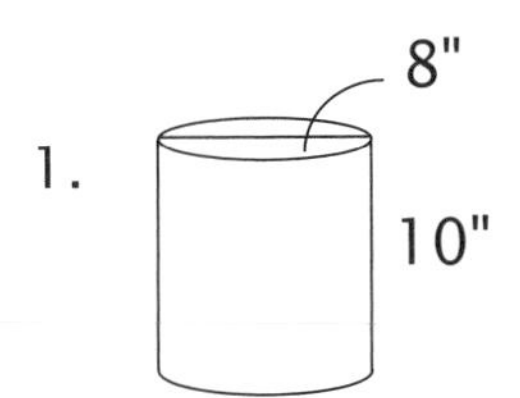

V = ____________

2. 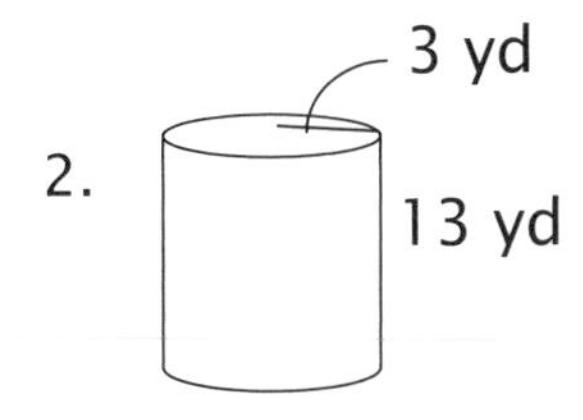

V = ____________

3. 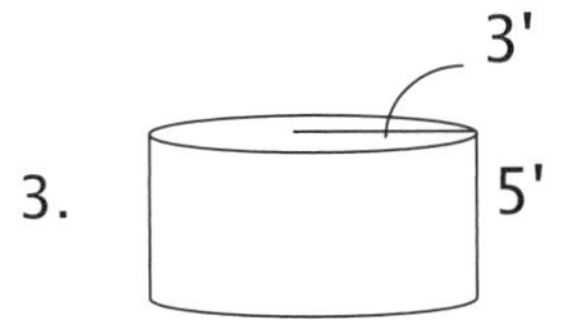

V = ____________

4. 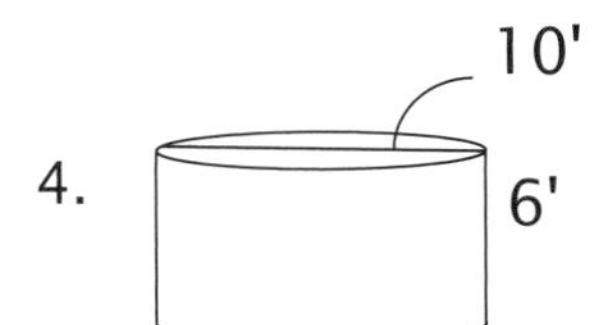

V = ____________

Add the polynomials.

5. $$\begin{array}{r} 5X^2 + 4X + 9 \\ \underline{+3X^2 + 8X + 1} \end{array}$$

6. $$\begin{array}{r} 8X^2 + 9X + 6 \\ \underline{-5X^2 - 3X - 2} \end{array}$$

7. $$\begin{array}{r} 3X^2 + 7X + 8 \\ \underline{-2X^2 + 3X - 9} \end{array}$$

8. The LCM of 10 and 15 = ____.

9. The GCF of 12 and 48 = ____ .

Change each percent to a fraction, then a whole number or a reduced mixed number.

10. 800% = —— = ____

11. 375% = —— = ____

12. 240% = —— = ____

Fill in the blanks with the correct geometric term (#13–16).

13. A figure with one dimension and definite starting and ending points

    is a ______________ .

14. A figure with two dimensions (both length and width) is a _________ .

15. A figure with no length or width is a _________ .

16. A figure with one dimension (infinite length but no width)

    is a _________.

17. What is normal body temperature in Celsius?

18. Which is the better deal, one-third off or 25% off?

19. One inch on a map represents 50 miles. What distance is represented by 3 1/2 inches?

20. On his test, Patrick got 18 out of 20 questions right. Write his score as a fraction and as a percent.

TEST

25

Build the rectangle and find the area or product.

1. $(X + 3)(X + 1) =$

2. $(X + 7)(X + 2) =$

3. $(2X + 1)(X + 3) =$

Multiply the binomials.

4. $\begin{array}{r} 2X+3 \\ \times\ X+4 \\ \hline \end{array}$

5. $\begin{array}{r} X+5 \\ \times\ X+7 \\ \hline \end{array}$

6. $\begin{array}{r} 2X+2 \\ \times\ X+5 \\ \hline \end{array}$

Add the polynomials.

7. $\begin{array}{r} 5X^2+4X+1 \\ +\ 8X^2+4X-2 \\ \hline \end{array}$

8. $\begin{array}{r} 3X^2-\ X-8 \\ +\ 7X^2-9X-1 \\ \hline \end{array}$

9. $\begin{array}{r} -4X^2+7X-6 \\ +\ 9X^2-3X+5 \\ \hline \end{array}$

Use the distributive property to simplify.

10. $8(4 + 45) =$

11. $6(Q + 3) =$

12. $X(X + 11) =$

Use order of operations to simplify.

13. $6^2 + 3 \cdot 5 - 4 + |-1| =$

14. $|20 \div 5 + (5)(6) - 3| =$

15. $(4^2 - 8) + 14 - 7 =$

Write a proportion for each set of similar polygons and solve for the unknown side.

16.

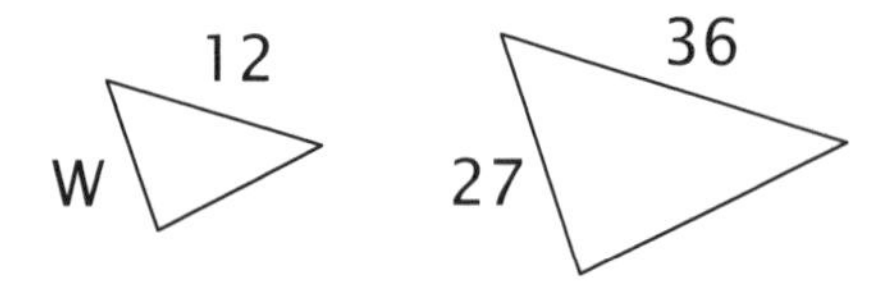

17.

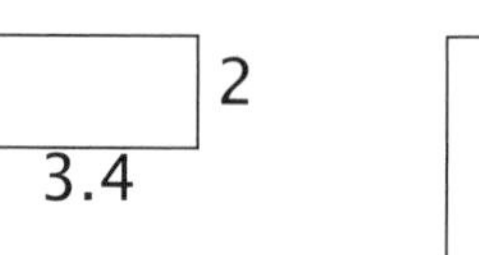

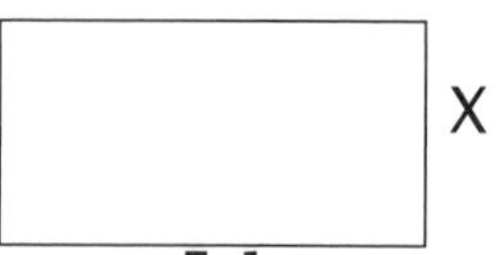

18. A coat that was originally $80.00 is now 40% off. Sales tax is 6%.
Is $50.00 enough to pay for the coat?

19. What geometric figure is formed by two rays extending from the same endpoint?

20. Which does not have infinite length: a line, a ray, or a line segment?

Add or subtract the times.

1. $\begin{array}{r} 4:32 \\ -\ 1:17 \\ \hline \end{array}$

2. $\begin{array}{r} 5:48 \\ -\ 2:50 \\ \hline \end{array}$

3. $\begin{array}{r} 4:16 \\ +\ 2:50 \\ \hline \end{array}$

4. $\begin{array}{r} 3:29 \\ +\ 2:48 \\ \hline \end{array}$

5. $\begin{array}{r} 6:37 \\ +\ 5:29 \\ \hline \end{array}$

6. $\begin{array}{r} 10:10 \\ -\ 2:36 \\ \hline \end{array}$

Multiply the binomials.

7. $\begin{array}{r} X+7 \\ \times\ X+8 \\ \hline \end{array}$

8. $\begin{array}{r} X+1 \\ \times\ X+3 \\ \hline \end{array}$

9. $\begin{array}{r} 2X+2 \\ \times\ X+6 \\ \hline \end{array}$

Write each decimal as a reduced fraction.

10. .8

11. .45

12. .70

Change each fraction to a decimal. Round your answers to the nearest hundredth.

13. $\frac{9}{10} =$

14. $\frac{1}{6} =$

15. $\frac{5}{7} =$

16. What name is given to an angle with a measure of 108°?

17. What name is given to an angle with a measure of 25°?

18. How many degrees are there in a right angle?

19. How many degrees are there in a straight angle?

20. Josh's plane took off at 6:15 and flew for 5 hours and 40 minutes. What time did his watch say when the plane landed?

Find the volume. Round to hundredths.

1.

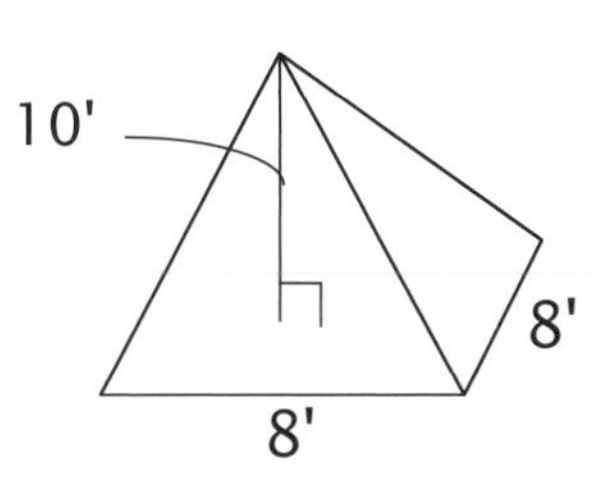

V = ____________

2.

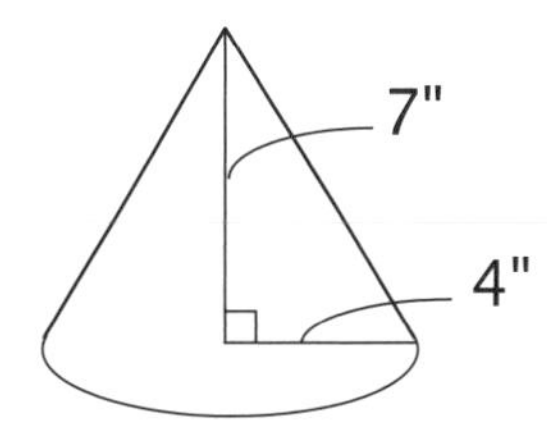

V = ____________

Add or subtract the times.

3. $\begin{array}{r} 4:28 \\ +\ 6:33 \\ \hline \end{array}$

4. $\begin{array}{r} 6:43 \\ -\ 2:51 \\ \hline \end{array}$

5. $\begin{array}{r} 7:35 \\ +\ 2:30 \\ \hline \end{array}$

Multiply to find the area or product. These may be built or set up as multiplication problems.

6. $(X + 6)(X + 6) =$

7. $(2X + 4)(X + 1) =$

8. $(A)(A + 8) =$

Follow the signs.

9. $\frac{7}{8} - \frac{1}{4} =$

10. $\frac{3}{4} \times \frac{5}{18} =$

11. $\frac{4}{5} \div \frac{1}{2} =$

Fill in the blanks (#12–16).

12. The number that comes most often in a list of data is called the ______________.

13. When a list of numbers is arranged in order from smallest to largest, the number in the middle is called the ______________.

14. The ______________ is another name for average.

15. An angle with a measure less than 90° is called an ______________ angle.

16. A point has no ______________ or ______________.

17. What is the square root of 36?

18. Give the mean of the following list of numbers: 21, 36, 42

19. How many degrees are there in a straight angle?

20. Each side of a square is $X + 5$ units long. What is the area of the square?

TEST

28

Change to military time.

1. 8:11 p.m.

2. 2:00 a.m.

3. 12:00 midnight

4. 5:30 a.m.

Change to standard time.

5. 1536

6. 2045

7. 0305

8. 1200

Add or subtract the military times.

9. $\begin{array}{r} 0550 \\ +\ 1730 \\ \hline \end{array}$

10. $\begin{array}{r} 2030 \\ +\ 0145 \\ \hline \end{array}$

11. $\begin{array}{r} 1619 \\ -\ 1128 \\ \hline \end{array}$

12. $\begin{array}{r} 0643 \\ -\ 0350 \\ \hline \end{array}$

Find the volume. Round to hundredths.

13. 4" 2.3" 2.3"

14. 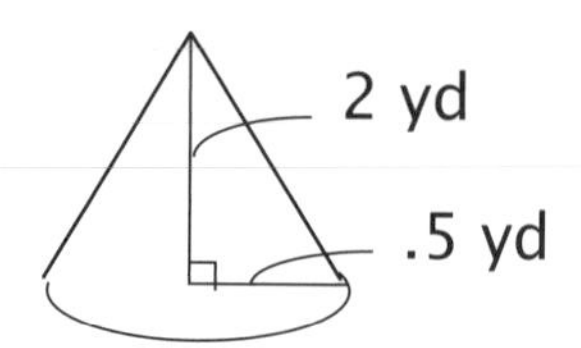

15. 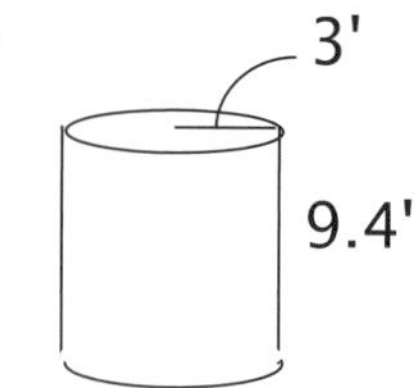

16. Alexis put five entries in the drawing for a trip to Florida. If there were 5,000 entries in all, what was the probability of Alexis winning the trip?

17. What is the GCF of 21 and 49?

18. Find the mode for the following list of numbers: 2, 2, 3, 4, 4, 4, 5, 6

19. A soldier begins his guard duty at 1230 hours. If his duty lasts 2 hours and 45 minutes, what time will it be when he comes off duty?

20. Briana got 24 questions right out of a possible 32. What was her percentage of right answers?

Add the measurements.

1. $\begin{array}{r} 29'\ 6" \\ +\ 5'\ 5" \\ \hline \end{array}$

2. $\begin{array}{r} 4 \text{ yd } 2 \text{ ft} \\ +\ 8 \text{ yd } 2 \text{ ft} \\ \hline \end{array}$

3. $\begin{array}{r} 9 \text{ lb } 10 \text{ oz} \\ +\ 6 \text{ lb } 11 \text{ oz} \\ \hline \end{array}$

Subtract the measurements.

4. $\begin{array}{r} 47'\ 3" \\ -\ 18'\ 4" \\ \hline \end{array}$

5. $\begin{array}{r} 5 \text{ yd} \qquad \\ -\ 2 \text{ yd } 2 \text{ ft} \\ \hline \end{array}$

6. $\begin{array}{r} 7 \text{ lb } 9 \text{ oz} \\ -\ 1 \text{ lb } 13 \text{ oz} \\ \hline \end{array}$

Change to military time.

7. 1:13 a.m.

8. 2:45 p.m.

Change to standard time.

9. 1030

10. 2107

Follow the signs.

11. $3\frac{5}{8} \div 1\frac{1}{4} =$

12. $5\frac{7}{9} \times 4\frac{2}{3} =$

13. $6\frac{1}{7} + 3\frac{3}{14} =$

14. What term is used for 1,000 grams?

15. What do we call 1/100 of a meter?

16. What part of a liter is a milliliter?

17. Mitch drew a name for the Christmas gift exchange. There were 55 names in the dish. If there are four people in Mitch's family, including himself, what is the probability of his drawing the name of someone in his family?

18. Ashton made the following scores on her spelling tests: 95, 85, 80, 90, 100. What was her mean, or average, score for that period of time?

19. What was Ashton's median spelling score? (See #18.)

20. What is the volume of a cone with a radius of 10 feet and a height of 10 feet?

TEST

30

Tell if the numbers are rational or irrational.

1. -15
2. $\sqrt{16}$
3. $\sqrt{5}$
4. π
5. 25
6. -7

Add or subtract the measurements.

7.
```
   8' 3"
 - 5' 5"
```

8.
```
   6 yd  1 ft
 + 4 yd  2 ft
```

9.
```
   19 lb   2 oz
 -  7 lb  11 oz
```

Use the scale to help you convert one metric measure to another.

kilo	hecto	deka	(unit)	.	deci	centi	milli
1,000	100	10	1		$\frac{1}{10}$	$\frac{1}{100}$	$\frac{1}{1,000}$

10. 10 centimeters = ______ millimeters

11. ______ kilometers = 23 meters

12. 5 liters = ______ centiliters

13. ______ grams = 2,000 milligrams

14. Are all rational and irrational numbers real?

15. Simplify and write $|-(4^2)|$ without the absolute value signs.

16. What is the name for an angle with a measurement of 175°?

17. What is the volume of a cylinder with a radius of 7 inches and a height of 5 inches?

18. What is the probability of drawing a black button from a box that holds 250 white buttons and 50 black buttons? (Be sure to find the total number of buttons first.)

19. Write .692 as a reduced fraction.

20. Give the mode for the following list of data: 21, 32, 33, 33, 76.

UNIT TEST **Lessons 23-30**

Add the polynomials.

1. $$\begin{array}{r} 4X^2 + 3X + 1 \\ +\ 7X^2 + 2X - 3 \\ \hline \end{array}$$

2. $$\begin{array}{r} 2X^2 - X - 6 \\ +\ 8X^2 - 1X - 2 \\ \hline \end{array}$$

3. $$\begin{array}{r} -5X^2 + 8X - 7 \\ +\ 6X^2 - 4X + 6 \\ \hline \end{array}$$

Multiply the binomials.

4. $$\begin{array}{r} 3X + 4 \\ \times\ \ X + 5 \\ \hline \end{array}$$

5. $$\begin{array}{r} X + 6 \\ \times\ X + 8 \\ \hline \end{array}$$

6. $$\begin{array}{r} 2X + 3 \\ \times\ \ X + 6 \\ \hline \end{array}$$

7. $(X + 2)(X + 1) =$

8. $(X + 6)(X + 4) =$

9. $(2X + 2)(X + 2) =$

Find the volume. Round to hundredths.

10. 11. 12.

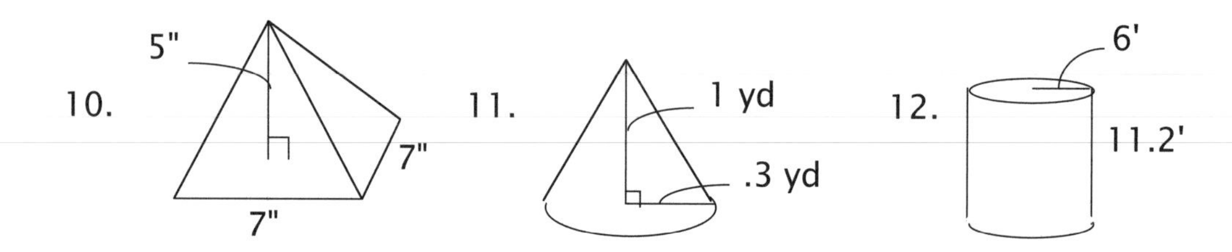

Add or subtract the times.

13. 3:21 − 1:06

14. 4:37 − 1:49

15. 5:28 + 7:52

Change to military time.

16. 10:19 p.m.

17. 4:00 a.m.

18. 5:30 p.m.

Change to standard time.

19. 0017

20. 0330

21. 1945

UNIT TEST

Add or subtract the military times.

22. $\begin{array}{r} 0540 \\ +\ 1720 \\ \hline \end{array}$

23. $\begin{array}{r} 2010 \\ +\ 0235 \\ \hline \end{array}$

24. $\begin{array}{r} 1012 \\ -\ 0642 \\ \hline \end{array}$

Add or subtract the measurements.

25. $\begin{array}{r} 9'\ 2'' \\ -\ 4'\ 6'' \\ \hline \end{array}$

26. $\begin{array}{r} 5\text{ yd}\ \ 2\text{ ft} \\ +\ 8\text{ yd}\ \ 2\text{ ft} \\ \hline \end{array}$

27. $\begin{array}{r} 12\text{ lb}\ \ 4\text{ oz} \\ -\ 5\text{ lb}\ \ 7\text{ oz} \\ \hline \end{array}$

Tell if the numbers are rational or irrational.

28. π

29. $\sqrt{25}$

30. $\sqrt{2}$

31. What is 250% of 34?

32. How many degrees are there in a right angle?

33. Find the mean of the list of data: 4, 6, 6, 7, 12

34. Meredith entered the drawing six times. If there are a total of 432 entries, what is the probability of one of her entries being drawn?

35. What is the name given to 1,000 meters?

FINAL EXAM

Follow the signs.

1. $(-8) + (-25) =$

2. $(-7) \times (-15) =$

3. $(11) - (-6) =$

4. $(-45) \div (9) =$

Simplify.

5. $-1^3 =$

6. $-(5)^2 =$

7. $(-8)^2 =$

8. $\left(-\frac{2}{3}\right)^2 =$

Write in exponential notation.

9. 95.214

Write in standard notation.

10. $1 \times 10^3 + 8 \times 10^2 + 2 \times 10^1 + 5 \times 10^0 + 6 \times \frac{1}{10^1}$

Simplify each expression.

11. $\sqrt{100} =$

12. $\sqrt{Y^2} =$

Simplify and solve for the unknown. Use order of operations as needed. Check your work.

13. $8 \cdot 2 + 5^2 - Y = 2(Y + 1) + 6$

14. Check

15. $8M - 4M - 6 - 3 + 5M = 8^2 - 1$

16. Check

17. $(-3)^2 \div 9 + 6 = D$

18. Check

Solve for the unknown.

19. $\frac{1}{8} = \frac{7}{Y}$

20. $\frac{11}{12} = \frac{A}{48}$

Write a proportion for each set of similar polygons and solve for the unknown side.

21. 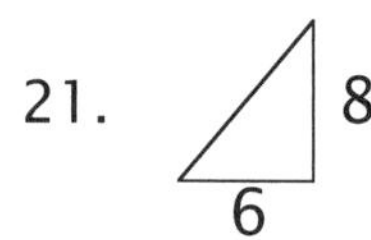

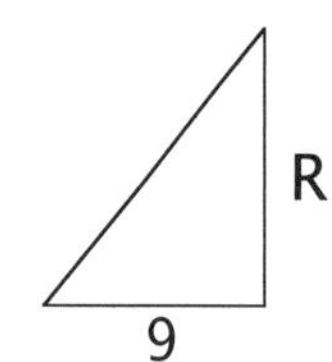

Find the LCM for each pair of numbers.

22. 3 and 4

23. 6 and 9

Find the GCF for each pair of numbers.

24. 24 and 40

25. 15 and 35

Add the polynomials.

26.
$$\begin{array}{r} 5X^2 + 4X + 2 \\ +\ 8X^2 + 3X - 4 \\ \hline \end{array}$$

27.
$$\begin{array}{r} 7X^2 - \ X - 3 \\ +\ 6X^2 - 2X - 5 \\ \hline \end{array}$$

28.
$$\begin{array}{r} -4X^2 + 9X - 8 \\ +\ \ 2X^2 - 6X + 1 \\ \hline \end{array}$$

29. What is the surface area of a rectangular solid with a length of seven inches, a width of five inches, and a height of six inches?

30. What is the surface area of a pyramid with a square base measuring 9 feet on a side if the slant height of each face is 11 feet?

31. The two legs of a right triangle measure 9 feet and 12 feet. What is the length of the hypotenuse of the triangle?

Multiply the binomials.

32. $\begin{array}{r} 2X+1 \\ +\quad X+6 \\ \hline \end{array}$

33. $\begin{array}{r} X+7 \\ +\quad X+9 \\ \hline \end{array}$

34. $\begin{array}{r} 2X+4 \\ +\quad X+5 \\ \hline \end{array}$

Find the volume. Round to hundredths.

35. 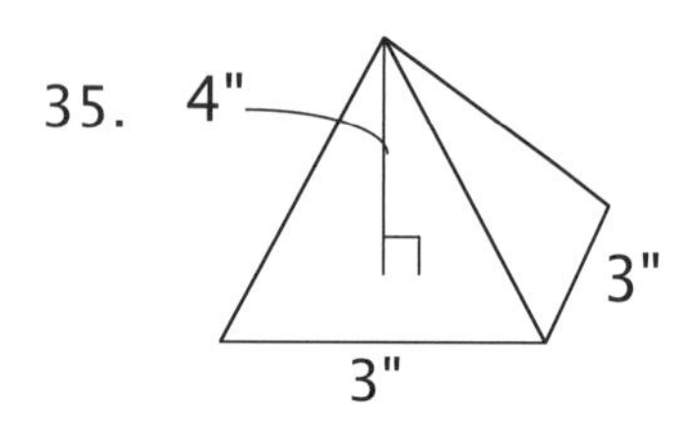

36. 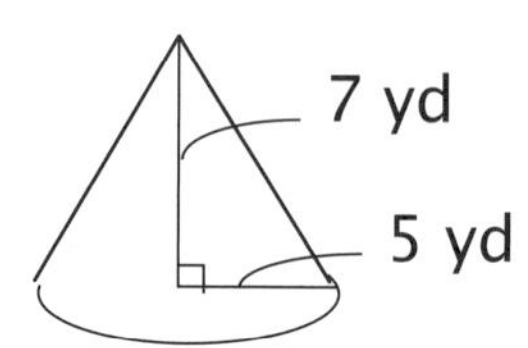

37. 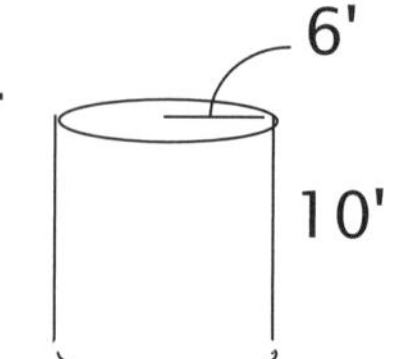

Add or subtract the times.

38. $\begin{array}{r} 7:18 \\ -\ 3:05 \\ \hline \end{array}$

39. $\begin{array}{r} 2:44 \\ +\ 1:59 \\ \hline \end{array}$

40. $\begin{array}{r} 0136 \\ +\ 0438 \\ \hline \end{array}$

Change to standard time.

41. 2120

42. 1611

43. 0345

Add or subtract the measurements.

44. $\begin{array}{r} 8'\ 3'' \\ -\ 5'\ 5'' \\ \hline \end{array}$

45. $\begin{array}{r} 6\text{ yd } 1\text{ ft} \\ +\ 9\text{ yd } 1\text{ ft} \\ \hline \end{array}$

46. $\begin{array}{r} 25\text{ lb } 8\text{ oz} \\ -\ 15\text{ lb } 10\text{ oz} \\ \hline \end{array}$

Tell if the numbers are rational or irrational.

47. π

48. $\sqrt{25}$